内容提要

北京服装学院服装艺术设计专业2012届优秀毕业设计作品集,凝结了211设计学专业四年的精髓设计。随着北京服装学院成为国际时尚设计联盟的领头羊,2012届北服学子们的设计作品更加令人瞩目,与时尚流行接轨的内容、色彩、图案、面料、廓形等等加丰富多彩,彼等秉北服在国内时尚系统具领跑地位的东风,在国际时尚国际流行设计院校行列的先锋地位。

图书在版编目(CIP)数据

北京服装学院服装艺术设计专业2012届毕业生作品集 / 北京服装学院编. —北京:中国纺织出版社,2012.6
 ISBN 978-7-5064-8719-1

Ⅰ.①北… Ⅱ.①北… Ⅲ.①服装设计-作品集-中国-现代 Ⅳ.①TS941.2

中国版本图书馆 CIP 数据核字(2012)第115839号

策划编辑:刘 茸 金 昊 责任校对:陈 红
责任设计:何 建 责任印制:何 艳

中国纺织出版社出版发行
地址:北京东直门南大街 6 号 邮政编码:100027
邮购电话:010—64168110 传真:010—64168231
http://www.c‑textilep.com
E‑mail:faxing @ c‑textilep.com
北京通彩色印刷有限公司印刷 各地新华书店经销
2012 年 6 月第 1 版第 1 次印刷
开本:787×1092 1/8 印张:17.5 插页:12
字数:100 千字 定价:198.00 元

凡购本书,如有缺页、倒页、脱页,由本社图书营销中心调换

北京服装学院
服装艺术与工程学院
2012届毕业设计作品集

GRADUATION WORKS 2012
FASHION COLLECTION BIFT

中国纺织出版社

1

2

开篇语

你现在翻开的这本作品集，便是我。
是我的梦想，是我的热爱，是我的理念，是我的技术，是我的汗水，是我的欣慰。
是我未来的开始。

这一段你给予我的时光中，
你教给我太多——你的严谨，你的创新，你的开放，你的鼓励。
近百名勤恳专业的老师，百余间设备齐全的教室，近百家实习试验基地。
你是我自始至终的坚实后盾。

所以未来的路上，
我要成为你的骄傲。
你的历史，我的传承；你的文化，我的创新；你的发展，我的努力。
前方或是成功、或是困难，我会执着，我有信心，
因为无论纽约、巴黎、米兰、伦敦、北京，我的梦想是要世界瞩目东方升起的光。

Preface

The portfolio you are reading now is mine.
It is my dream, my love, my concept, my skills, my hard work, and my pleasure.
It is the beginning of my future.

In this period of time,
You have taught me so much,
Your precision, your creativity, your open mind, and your ambition.
Hundreds of professional educators, hundreds of well-equipped classrooms, internship exercitation bases,
They are my solid backing from beginning to end.

Therefore in the future,
I want to be your pride,
Your history is my inheritance; your culture my creativity; your development my hard work.
There may be success and may be difficulty, but I will always have persistence and faith.
Because no matter where I am, New York, Paris, Milan, London, Beijing,
My dream is to let the world see the light that is rising from the east.

艺术寄语

北京服装学院艺术设计学本科 2012 届毕业生迎来了人生中一个非常重要的日子，那是北京服装学院艺术设计学本科的第一届毕业生的毕业作品汇报展。这次毕业生作品汇报展是以"北京服装学院中国国际时装周"、"中国 MODART"、2012 北京·SEVEN DAYS"、"北京设计周"的形式呈现。这充分展示了北京服装学院艺术设计人才的培养理念和教学水平，也蕴含了北京服装学院艺术设计学本科师生的辛勤劳动和无限智慧。中国有句老话，从小看大，从这些毕业生设计作品中我们看到了中国服装业的美好未来。

作为北服的院长，我感到非常欣慰，看着我们的作品不断走向成熟，每当我们的学生一步步成长，我为了他们的每一次进步，为了他们设计作品打动自己和感动大家，为我所有真心付出的老师们——感谢你们，同志们，你们永远是我们的骄傲代表，为你们的付出感到骄傲和自豪！

当今设计的意义不仅仅是强调艺术文化的内涵、中国精神的展现。中国服装艺术设计的根基在中国，在中国无比广袤的土地上根植于中国国际时装周上每个人的心。设计的根基，就是要与人互动沟通和为人民服务的设计和创新内涵。北京服装学院艺术设计学本科在推动发展的过程中，逐渐形成了独树一帜的特点。在近几年内，北京服装学院所形成的十几个工作室和3家工作室的目前成果，是我们努力的方向。人才的队伍建设与教学体系建设不断扩大，北京服装学院全体教师对北服的付出，为学生的成长做出的贡献。在此，我代表北京服装学院向他们致以崇高的敬意！

当好21世纪的接班人，为人、为学、为艺的态度，是对"四方面的责任"的代代相传，我们一代代服装设计和艺术学的中国服装艺术设计与中国的未来，为向中国国际时装周的发展——祝愿。 过去的日子里，伴随着中国国际时装周的发展大踏步，北京服装学院各学科各专业紧大步，已能跻身于中国设计艺术教育的先进行列，该校为我国的ITAA国际时装周、亚洲服装等各日益辉煌。如今的一个多月天，同时间也经过北京建设文化交流中的中国服装界一次又一次走进北京、走进北京服装学院亚洲时装周、海内外国际时装展。国际化人才的培育方式在，北京服装学院教师的共同努力下，北京服装学院2的设计师团队，每每成为一个从北京走出的服装设计大名学和设计师长和培养出来的设计师，形成了北京服装学院的自信和自豪，也鼓励我们的大家为师生走向世界并创造出北京服装的成长未来！

刘元风
北京服装学院院长

Foreword of President of BIFT

This portfolio is 2012 BIFT graduation collection. These works are designed by the students majored in Fashion Art and Design of BIFT. There are only a small part of graduation designs in the portfolio, but it still can reflect how BIFT has been innovating and making constant progress. These designs titled by "BIFT • SEVEN DAYS Graduation Collection" were presented in a joint fashion show with Shih Chien University and MOD'ART International in 2012 China Spring Fashion Week and won high appraisal by the apparel industry and media, who were deeply impressed by the innovation and design quality, and commented these works provided them with more expectations on the future of China's fashion industry.

As the president, I am very proud of these students. Our teachers and students have been working very hard because they are always trying to improve and innovate. They are making great efforts to pursue their dreams and honors. I would like to congratulate them and show my respect to them.

Design nowadays is in the high-speed channel of technology development and the concept of innovation. Chinese art and design education is evolving new strategies to cultivate students' ability of adapting to China apparel industry and the increasing international competition. The cooperation between enterprises and universities, and the interaction between business community and academia are becoming the critical aspects of the education reform and innovation. In the recent five years, BIFT has established more than 100 internship bases with enterprises. Over ten scholarship have been sponsored. There are three research centers funded by a ten-year contract. Furthermore, many enterprises provide materials, equipments, technology and funding support for the graduation collection. Please allow me on behalf of BIFT to express high respect and sincere thanks to all the friends, enterprises, and designers.

We are facing complex global economy, industry and market changes in the 21st century. China apparel industry has been making rapid progress though we just started a few decades ago. The focus is being transferred from west to east. Nevertheless, international communication, integration, and interconnection become a certain trend. In the China Spring Fashion Week, BIFT invited Shih Chien University and MOD'ART International to present a joint fashion show. At the same time, we held ITAA international academic conference. The development of BIFT to some extend represents the development and innovation of China fashion design education. All these efforts would definitely promote the apparel industry and creative industry in China. Moreover, this will further motivate Beijing to be a city of culture, creation, and fashion. I believe that BIFT will be China's leading and internationally renowned design university through our constant efforts. I hope that all the graduates will contribute to the fashion and apparel industry by what they have studied, experienced, and trained in BIFT.

President of Beijing Institute of Fashion Technology

Liu Yuanfeng

目录
Contents

Section 1 返乡 / Reverse / p19

Section 2 融化的冰山 / Melting Iceberg / p34

Section 3 绽放 / Bloom / p54

Section 4 凝聚 / Condensation / p70

Section 5 很正经 / So Serious / p99

Section 6 不羁 / Unruly Hippie / p109

Section 7 流浪 / Life is Tramp / p122

Section 8 边缘 / The Edge / p128

工作人员名单 / Staff List / p136

"北漂·SEVENDAYS" 北漂北京设计竞赛获奖名单及作品名称 / p137

鸣谢 / Thanks / p137

模特 / Models / p138

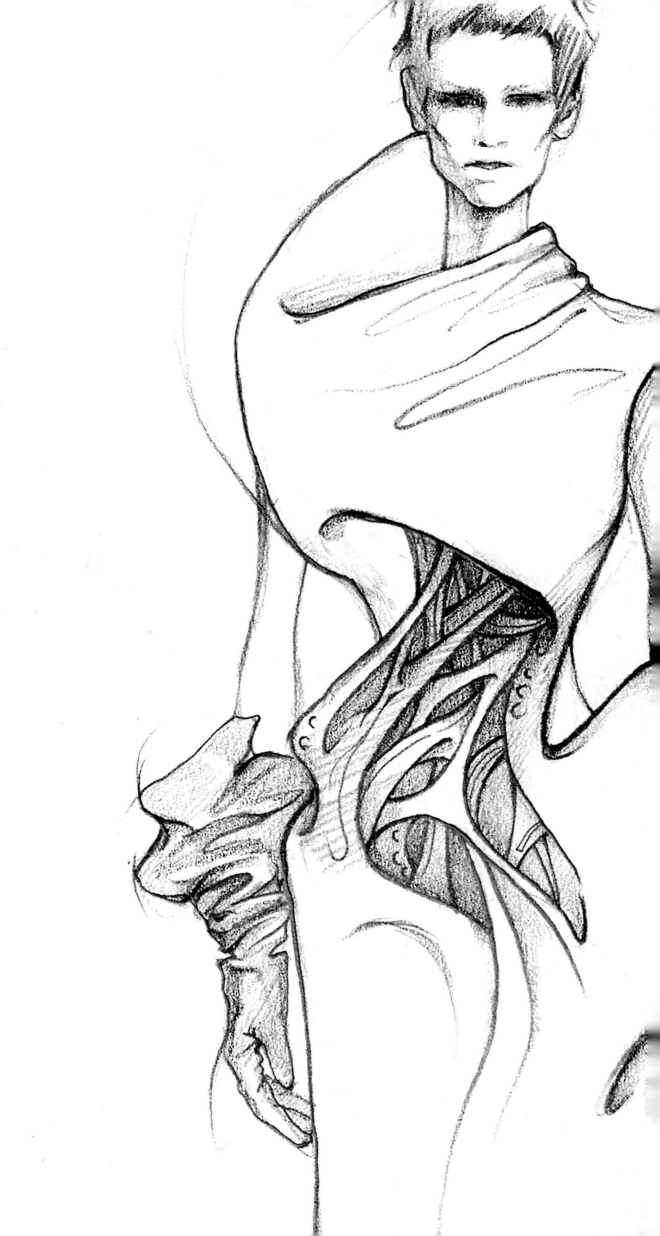

Section 1
逆向 / Reverse

设计师 / Designer
周讵燕 / Zhou Juyan

"汉帛奖"第20届中国国际青年设计师时装作品大赛 金奖
Golden Award HEMPEL AWARD the 20th International Young Fashion Designers Contest

指导教师 / Design Advisor
苏步 / Su Bu

模特 / Model
王梦雅 / Wang Mengya

设计师 / Designer
周诅燕 / Zhou Juyan

"汉帛奖"第20届中国国际青年设计师时装作品大赛 金奖
Golden Award HEMPEL AWARD the 20th International Young Fashion Designers Contest

指导教师 / Design Advisor
苏步 / Su Bu

模特 / Model
李蔚语 / Li Weiyu

设计师 / Designer
周讵燕 / Zhou Juyan

"汉帛奖"第 20 届中国国际青年设计师时装作品大赛 金奖
Golden Award HEMPEL AWARD the 20th International Young Fashion Designers Contest

指导教师 / Design Advisor
苏步 / Su Bu

模特 / Model
赵冰清 / Zhao Bingqing

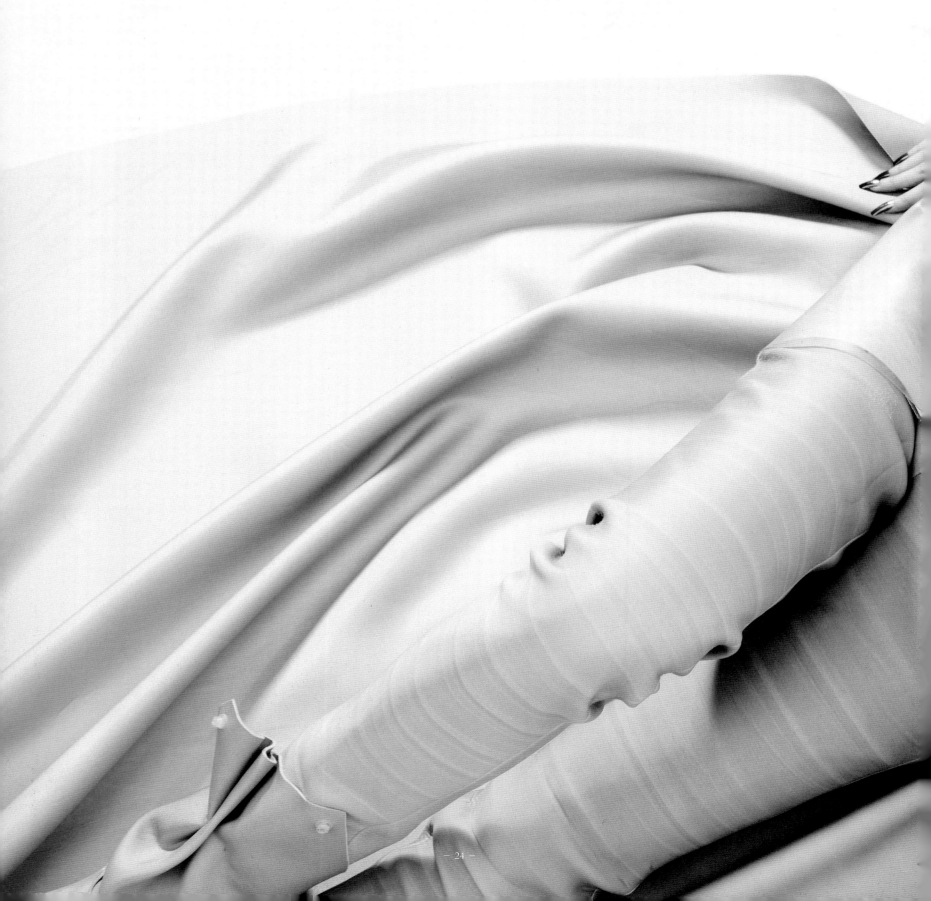

设计师 / Designer
周讵燕 / Zhou Juyan

"汉帛奖"第20届中国国际青年设计师时装作品大赛 金奖
Golden Award HEMPEL AWARD the 20th International Young Fashion Designers Contest

指导教师 / Design Advisor
苏步 / Su Bu

模特 / Model
李姿含 / Li Zihan

设计师 / Designer
周讵燕 / Zhou Juyan

"汉帛奖"第20届中国国际青年设计师时装作品大赛 金奖
Golden Award HEMPEL AWARD the 20th International Young Fashion Designers Contest

指导教师 / Design Advisor
苏步 / Su Bu

设计师 / Designer
李佳佩 / Li Jiapei

SEVEN DAYS 2012 "北服杯" 银奖
Silver Award SEVEN DAYS BIFT CUP 2012

指导教师 / Design Advisor
谢平 / Xie Ping

模特 / Model
徐文晞 / Xu Wenxi   王梦雅 / Wang Mengya
贾亦真 / Jia Yizhen   李梦涵 / Li Menghan   孙婷 / Sun Ting

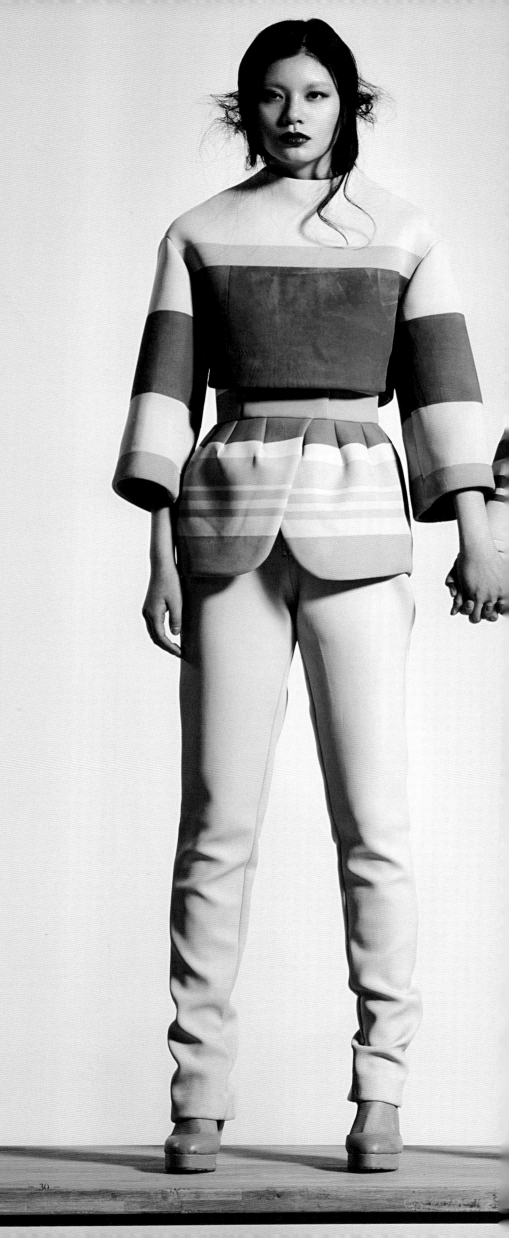

设计师 / Designer
李佳佩 / Li Jiapei
SEVEN DAYS 2012 "北服杯" 银奖
Silver Award SEVEN DAYS BIFT CUP 2012

指导教师 / Design Advisor
谢平 / Xie Ping

模特 / Model
王梦雅 / Wang Mengya

Section 2
融化的冰山 / Melting Iceberg

设计师 / Designer
张晓田 / Zhang Xiaotian
SEVEN DAYS 2012 "北服杯" 金奖
Golden Award SEVEN DAYS BIFT CUP 2012

指导教师 / Design Advisor
孙雪飞 / Sun Xuefei

模特 / Model
徐文晞 / Xu Wenxi

设计师 / Designer
唐青 / Tang Qing   陈丽莎 / Chen Lisha

指导教师 / Design Advisor
兰岚 / Lan Lan

模特 / Model
李姿含 / Li Zihan   蔡浩 / Cai Hao

设计师 / Designer
唐青 / Tang Qing　陈丽莎 / Chen Lisha

指导教师 / Design Advisor
兰岚 / Lan Lan

模特 / Model
王玺皓 / Wang Xihao

设计师 / Designer
彭湘婕 / Peng Xiangjie

指导教师 / Design Advisor
苏步 / Su Bu

模特 / Model
李姿含 / Li Zihan

设计师 / Designer
彭湘婕 / Peng Xiangjie

指导教师 / Design Advisor
苏步 / Su Bu

模特 / Model
张路野 / Zhang Luye

设计师 / Designer
彭湘婕 / Peng Xiangjie

指导教师 / Design Advisor
苏步 / Su Bu

模特 / Model
唐晓天 / Tang Xiaotian

设计师 / Designer
彭湘婕 / Peng Xiangjie

指导教师 / Design Advisor
苏步 / Su Bu

模特 / Model
鞠鹏志 / Ju Pengzhi

设计师 / Designer
李胜男 / Li Shennan 张健 / Zhang Jian 杨雨心 / Yang Yuxin

指导教师 / Design Advisor
郭瑞萍 / Guo Ruiping

模特 / Model
徐文晰 / Xu Wenxi 李姿含 / Li Zihan

设计师 / Designer
胡文科 / Hu Wenke 胡玲群 / Hu Lingqun
王元 / Wang Yuan 吕茜 / Lv Xi

指导教师 / Design Advisor
邱佩娜 / Qiu Peina

模特 / Model
王玺晓 / Wang Xiliao 蔡浩 / Cai Hao

Designer 杨双 / Yang Shuang 马华远 / Ma Huayuan
Design Advisor 张正学 / Zhang Zhengxue

设计师 / Designer
胡文科 / Hu Wenke 胡羚群 / Hu Lingqun
王元 / Wang Yuan 吕熹 / Lv Xi

指导教师 / Design Advisor
邱佩娜 / Qiu Peina

模特 / Model
李博文 / Li Bowen 唐晓天 / Tang Xiaotian
张路野 / Zhang Luye

设计师 / Designer
朱晓文 / Zhu Xiaowen

指导老师 / Design Advisor
谢平 / Xie Ping

模特 / Model
赵晨池 / Zhao Chenchi   李沛霖 / Li Peilin

李克瑜
原中国服装设计师协会副会长
北京服装学院顾问

这次的毕业设计作品总体上来说是很成功的，学生们的思路很开阔，设计表达得也很好。这届作品在色彩的把握、图案的运用、廓型的塑造、面料的选择上都可以看出是经过这些年轻设计师们反复推敲和思索过的。学生们的思维很活跃，作品的风格很多，可以用百花齐放来形容。他们通过各种艺术手法将自己的思想表现到服装中去，也在积极地探求各种元素的沟通与揉合。

我想告诫年轻的设计师们：要坚持梦想，做事情要持之以恒，希望他们在以后的道路上把服装做好做精；再者，希望年轻的设计师们关注相关艺术学科的发展，触类旁通地运用到服装设计中去；另外，中国的传统文化是很值得探究的，希望年轻的设计师们把祖国的文化发扬光大，时刻牢记自己的文化根基，让中国传统时尚起来。

Li Keyu
Former Vice President of China Fashion Designer Association
Consultant of Beijing Institute of Fashion Technology

The BIFT graduation collection is successful. These students are open-minded and show good abilities of design presentation. The colors, patterns, silhouette and materials are worked out after repeated deliberation. They melt their own values and ideas with the fashion design through a variety of artistic techniques and are also actively exploring the various elements of the communication and blend.

I suggest that young designers should insist on their dreams and need to persevere, and pay more attention to the development of science and technology, and integrate these to their designs. Moreover, young designers should also study our own culture deeply and try to create unique design based on the culture. I do hope these designers can spread our culture by fashion to the whole world.

刘元风
中国服装设计师协会副主席
北京服装学院设计艺术分院院长
北京服装学院教授

我每年都参加北京服装设计师生作品发表会,今年的作品更具有新意,具有新的思想,有着浓厚的时尚气息与文化内涵。学生们的作品各有特色,今年的作品让我看到她们对于专业的热爱,设计主题与所用的面料都有新的突破,这是中国学生自己的作品。通过几个主题的划分,今年学生对色彩分别的运用,向我们展示了,她们独特的个人想法。所选面料,工艺技巧,以及材料都非常精细,这些都体现出未来所需要的,有一种令人震撼,新奇的感觉。所谓服装,一定是由构成美感的要素,我看到了别具一格,一定是在同样美感的元素,我看到明年市场方面将出的新品。

希望同学们进入社会后其他的建筑,继续努力学习,为自己今后接触到的品牌设计与什么样的品牌文化,为自己将来的品牌设计打下基础,做一个有远见的设计师。当然,这些也是未来所需要的,有一种令人震撼,新奇的感觉。与北京服装学院携手,共同自己的风采融入,了解市场,迎合市场的需要,融入市场方能立足于市场。

Liu Yuanfeng

Vice President of China Fashion Designer Association
Vice Director of Beijing Apparel and Textile Association
President of Beijing Insitute of Fashion Technology

Every year I attend the BIFT graduation collection as a judge. I feel that the works of this grade are more mature. These works convey their insight about fashion trend instead of playing meaningless concepts. I believe this comes from the hard work of teachers and students. Our teachers directed the students to combine their own ideas with the main design concepts, and present them by different themes. Moreover, diverse colors were used and looked nice as a whole Generally, the design of this year is better than that of last year. However, efforts are still needed, such as designing better accessory, which plays significant role in the fashion show. I expect it can be improved next year.

 I hope these students can be humble when they are starting their new career. They should adapt what they have studied in the university to company business strategy and products features. At the beginning, it is critical to understand the meaning of brands and markets instead of emphasizing too much on the personal style. How to work with a team is also important. To sum up, the sooner the students understand the fashion market and real-world business, the better they will contribute to the industry.

谢衣萍

香港演艺大学舞美设计系助理教授

我觉得他们作品的设计有国际水平，这一点是这次作品的最大特点。每一个人有自己所要表达的设计，不管是装饰，色彩配搭及材料，都是非常完整。他们所达到的效果，都做得非常仔细，所以他们自己主导着自己的结果。我们称这是有多一分对的存在。之后再向着他们的生活去发展。

我觉得未来中国也要从设计为主要的生存去做。因为这样他们的设计才能够发展。我而我们这些做评审者，他们可能是未来的作品，我一"哈，怎么这个这么有意思！"我觉得他们自己是可以这样玩玩的，就说他说这样设计出来，是互相激励的，这我可以继续下去。

Hsieh Grace
Assistant Professor of Shih Chien University, Taiwan

I feel really exciting about the BIFT graduation collection. Their works are definitely in the international level. I kept telling the judge beside me that I really like these works. Every detail from head to foot was elaborately designed. The apparels were just marvelous, and the series were nicely completed. Afterwards, the collection from Paris showed more flavor of ready-to-wear compared with the BIFT collection. I was deeply impressed by the material dyeing. I think the students in BIFT are very lucky to have so many opportunities and laboratories where they can experiment and create so wonderful works. Unfortunately, Shih Chien University does not have some of these laboratories.

I do hope two universities can promote more cooperations because many students from our university are largely inspired by the work of BIFT students. "Wow, we can do design in this way!" becomes the most common comments from students of Shih Chien University. I believe students from two universities can learn and inspire from each other. I hope this will come true.

丁伟
SEVEN DAYS 品牌 CEO

北服的毕业设计作品一年比一年好，我去年也是评委，今年的设计水准要高很多，整体蛮不错的。相信同学们毕业后无论是自己创业还是融入企业都有成功的可能性。当然，自己创业的学生一定是非常自主的，是一种纯粹的诉求。中国需要独立设计师，独立创业不单单代表自己经营品牌，也会为商家服务。另外一部分学生兼容性比较好，到企业里去锻炼和积累，是很好的实践。我觉得现在的年轻设计师作为中国时尚界的未来，必须着眼于未来市场的需求，而不是当下的市场。中国的时尚产业同工业一样落后西方几十年，这种落后的构成主要是三方面：一是商家的落后；二是渠道、模式的落后；三是市场的消费落后。这几年通过互联网的传播，国门更大地打开，这种局面在逐渐好转，那么同学们现在需要做的是在专业方面潜心去修炼，坚持下去，再关注媒体宣传等方面。待成熟的经营模式壮大后，好的商业模式就会带动他们向前。

Ding Wei

CEO of Seven Days Fashion Co.,Ltd.

BIFT student graduation collections are always great. Compared with the collection of last year, as one of the judges, I think the design quality is becoming better. I believe that there would be many opportunities for the students who either start their own business or find a job in a company. Independent students may prefer to establish and develop their brands and products. It however does not necessarily mean that you have to run your own label. Your business may provide certain source or service for other business entities. Working for others' company is also a very important experience for those students who are flexible and eager to learn. I think, young designers, obviously as the tomorrow of the apparel industry, must have the vision on the future market instead of the present. The creative and fashion industry in China still falls behind due to the developing level of fashion companies, the lack of innovation on channels and business patterns, and the immatureness of the market and consumers. Nevertheless, the situation is getting improved. I hope that the students can educate themselves in practice after graduation and keep eyes on propaganda, and remember a good business pattern will make you succeed.

吴宏坤
上海索雅时装有限公司董事总经理

这一次的毕业作品非常好，学生进步很大。学校的变化也很大，我是1999年离开学校的，以前学校的毕业作品展示没有做得这么好，现在学校集合了各方面的人力物力，学生的才华可以展示得这么淋漓尽致，我觉得现在这样非常好。

从市场化的角度来看，现在大家的设计可能偏重于艺术，实穿性少了一些，但是其实服装毕竟不是纯艺术，我认为它是商业艺术，我认为如果实用性和艺术性结合得再更好一些，那就更有市场了。

Wu Hongkun
General Manager of Shanghai Suoya Fashion Co.,Ltd.

The fashion collection this year is great. I can feel the students are making huge progress. I graduated in 1999 from BIFT. The show at that time was far simpler than now. The students nowadays have much better platform to demonstrate themselves and their designs. This is wonderful.

Their designs concentrate more on the aesthetics. A real good design is an art related to markets. I believe that it would be better to have both aspects integrated.

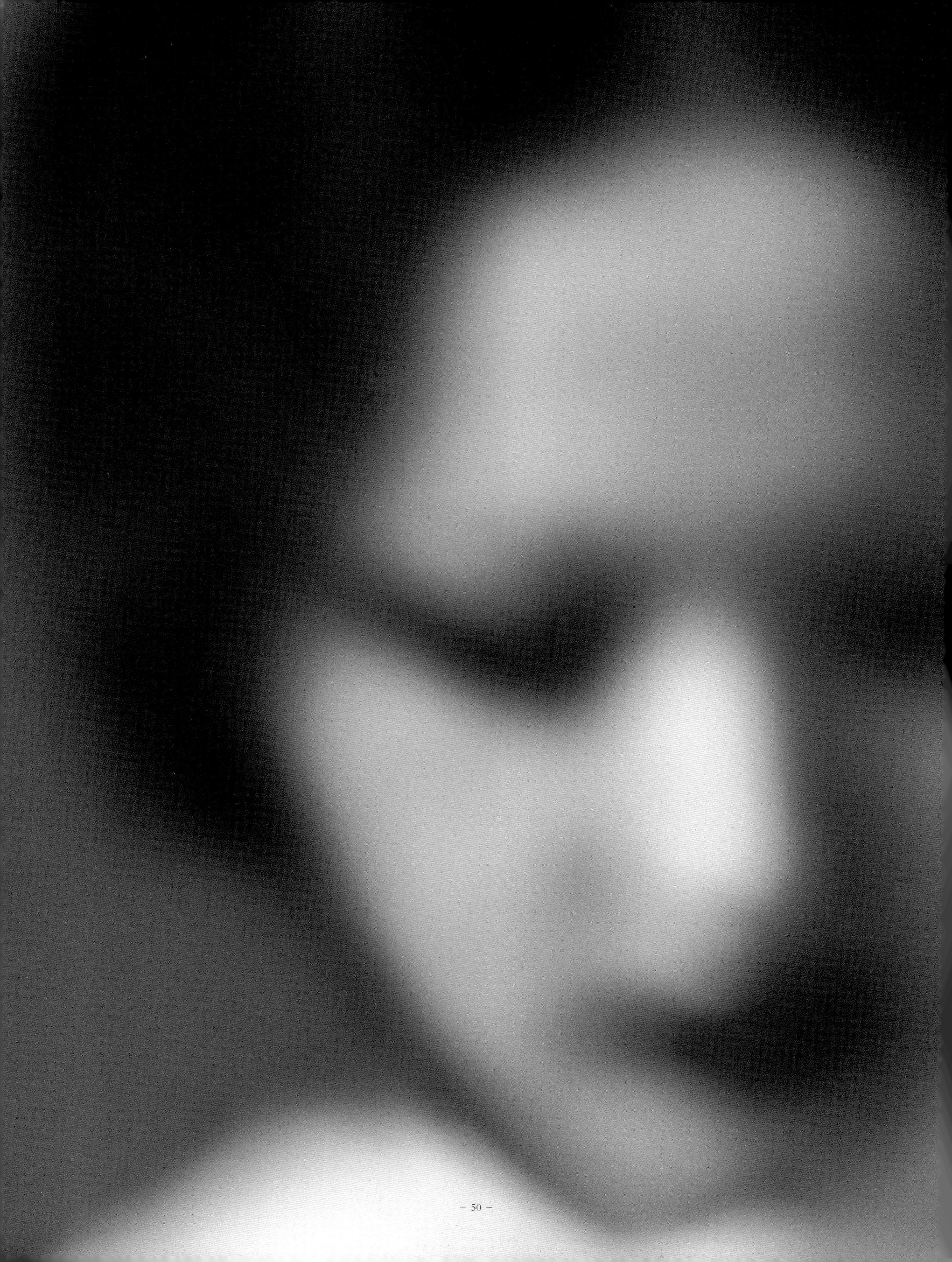

设计师 / Designer
朱晓文 / Zhu Xiaowen

指导教师 / Design Advisor
谢平 / Xie Ping

模特 / Model
赵晨池 / Zhao Chenchi   刘丽洁 / Liu Lijie

Section 3
绽放 / Bloom

Designer
秦泓鹏 / Qin Hongpeng 刘飞 / Liu Fei
尹哲 / Yin Zhe 赵岩 / Zhao Yan

Design Advisor
苏步 / Su Bu

Model
孙婷 / Sun Ting

设计师 / Designer
朱晓文 / Zhu Xiaowen

指导教师 / Design Advisor
谢平 / Xie Ping

模特 / Model
孙婷 / Sun Ting

设计师 / Designer
秦鸿鹏 / Qin Hongpeng　刘飞 / Liu Fei
尹哲 / Yin Zhe　赵研 / Zhao Yan

指导教师 / Design Advisor
苏步 / Su Bu

模特 / Model
贾亦真 / Jia Yizhen

设计师 / Designer
靳云英 / Jin Yunying　陈小姣 / Chen Xiaojiao
王雪桦 / Wang Xuehua　王晓旭 / Wang Xiaoxu

指导教师 / Design Advisor
杜冰冰 / Du Bingbing

模特 / Model
鞠鹏志 / Ju Pengzhi

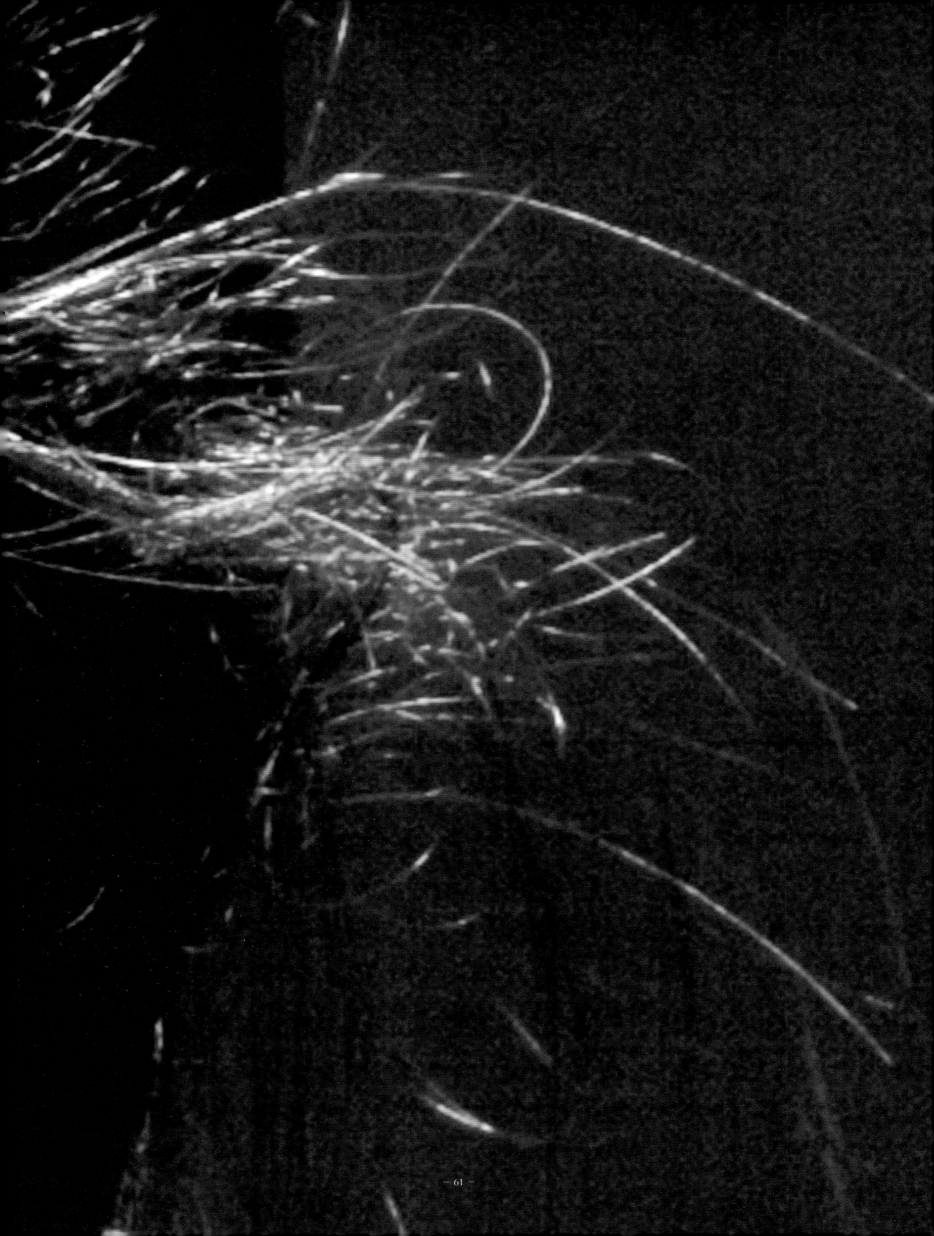

Designer
陈小姣 / Chen Xiaojiao　靳芸莹 / Jin Yunying　王雪华 / Wang Xuehua　王晓旭 / Wang Xiaoxu

Design Advisor
杜冰冰 / Du Bingbing

Model
钟璐淳 / Zhong Luchun

设计师 / Designer
靳云英 / Jin Yunying   陈小姣 / Chen Xiaojiao
王雪桦 / Wang Xuehua   王晓旭 / Wang Xiaoxu

指导教师 / Design Advisor
杜冰冰 / Du Bingbing

模特 / Model
刘丽洁 Liu Lijie

设计师 / Designer
秦洪鹏 / Qin Hongpeng 刘飞 / Liu Fei
尹哲 / Yin Zhe 赵岩 / Zhao Yan

指导教师 / Design Advisor
苏步 / Su Bu

模特 / Model
王梦雅 / Wang Mengya

设计师 / Designer
张晓丹 / Zhang Xiaodan

SEVEN DAYS 2012 "北服杯" 金奖
Golden Award SEVEN DAYS BIFT CUP 2012

指导教师 / Design Advisor
孙雪飞 / Sun Xuefei

模特 / Model
赵辰池 / Zhao Chenchi

设计师 / Designer
张晓田 / Zhang Xiaotian

SEVEN DAYS 2012 "北服杯" 金奖
Golden Award SEVEN DAYS BIFT CUP 2012

指导教师 / Design Advisor
孙雪飞 / Sun Xuefei

Section 4 凝 / Condensation

设计师 / Designer
胡文邦 / Hu Wenbang

"汉帛奖"第20届中国国际青年设计师时装作品大赛 银奖
Silver Award HEMPEL AWARD the 20th International Young Fashion Designers Contest
SEVEN DAYS 2012 "北服杯" 铜奖
Bronze Award SEVEN DAYS BIFT CUP 2012

指导教师 / Design Advisor
苏步 / Su Bu

模特 / Model
那广子 / Na Guangzi

设计师 / Designer
郭奇 / Guo Qi

指导教师 / Design Advisor
邱佩娜 / Qiu Peina

模特 / Model
谢腾 / Xie Teng

设计师 / Designer
胡文邦 / Hu Wenbang

"汉帛奖" 第 20 届中国国际青年设计师时装作品大赛 银奖
Silver Award HEMPEL AWARD the 20th International Young Fashion Designers Contest
SEVEN DAYS 2012 "北服杯" 铜奖
Bronze Award SEVEN DAYS BIFT CUP 2012

指导教师 / Design Advisor
苏步 / Su Bu

模特 / Model
谢腾 / Xie Teng

设计师 / Designer
胡文邦 / Hu Wenbang

"汉帛奖"第 20 届中国国际青年设计师时装作品大赛 银奖
Silver Award HEMPEL AWARD the 20th International Young Fashion Designers Contest

SEVEN DAYS 2012 "北服杯"铜奖
Bronze Award SEVEN DAYS BIFT CUP 2012

指导教师 / Design Advisor
苏步 / Su Bu

设计师 / Designer
胡文邦 / Hu Wenbang

"汉帛奖"第20届中国国际青年设计师时装作品大赛 银奖
Silver Award HEMPEL AWARD the 20th International Young Fashion Designers Contest

SEVEN DAYS 2012 "北服杯" 铜奖
Bronze Award SEVEN DAYS BIFT CUP 2012

指导教师 / Design Advisor
苏岩 / Su Ba

模特 / Model
靳亦真 / Jin Yizhen

设计师 / Designer
刘丽 / Liu Li 赵乃漩 / Zhao Naixuan 刘萌 / Liu Meng

指导教师 / Design Advisor
梁燕 / Liang Yan

模特 / Model
王昱超 / Wang Yuchao

娜塔莉·杜赛特
美国时装艺术基金会主席

我觉得这是场很棒、很有活力的秀。它给很多学生提供了展示自己的平台，也使我们看到了许多关于当代中国学生的想法，他们的想法是与我们经常见到的美国的或是其他国家的都不一样的，很独特。我认为对于中国年轻的设计师来说，保持中国的独特的文化和自己的创造力，并与在学校里学到的那些感兴趣的知识结合来做设计是很重要的。

我觉得你们一定要努力保持住自己的文化，围绕着你们自己的文化作一些独特的设计；不要看那些流行趋势，不要太商业化。因为现在有太多的成衣品牌，而我们更希望从你们这些年轻的设计师上看到的是与他们不同的设计，是有你们自己文化和思想的设计。就像20年前，日本的设计师将有自己独特文化的设计推向了世界，得到了世界的广泛认可，这也是我们希望在中国的设计师和作品上看到的未来，流淌着你们文化滋养的未来。

Nathalie Doucet

President of Arts of Fashion Foundation

I think it's great and genius. Some technology gives me a very strong impression, because the students keep the culture of China in mind. What we want to see is something really unique from what we often see in America. I think it's very important for the young Chinese designers to keep their culture and put what they like in school in their design and make it perfect.

Try to keep your culture. Try really doing something unique around your culture. Don't try to look at the trends. Don't try to be too much commercial because there are so many brands right now, so many garments. The only thing we want to see is something really unique that can be seen only in your culture. Like the Japanese 20 years ago, they found their own way to express their unique fashion by focus on their culture and it was a big success. This is what we want to see in China.

常青
北京纺织服装行业协会会长

首先对2012北服毕业生作品集的精彩呈现表示衷心的祝贺!
自从与服装业结缘,七、八年来,每年北服毕业生作品的发布会都成为我必看的时装秀,而且每次都让我感到格外亲切和欣喜。特别是近几年来,同学们的作品由原来的略带青涩和稚嫩变得更加成熟,那些充满创意而又制作精良的优秀作品,让人很难想象竟是出自尚未走出校门的学生之手!这让我们看到服装界的后继有人,看到产业的未来充满希望。
中国服装要从制造走向创造,核心是设计,关键是人才。你们正面临着极好的发展机遇,也肩负着重要的历史使命。要看到,服装设计师成长之路任重而道远。希望你们以今天的毕业为起点,在未来的职业实践中不断充实和提升自己,在文化素养的积累、综合审美的培养、与市场对接的历练等方面,做出不懈的努力,真正成为中国服装业新崛起的栋梁之才,为实现中华民族的伟大理想贡献力量!

Chang Qing
Director of Beijing Textile and Apparel Association

Congratulations on all the 2012 BIFT graduates and their collections!
Since I joined the apparel industry, BIFT graduation collection has become a must-see fashion show every year. It is amazing and enjoyable every time. Especially in recent years, students, works are becoming much more mature. It is hard for me to imagine how those outstanding, creative and well-made works are made just from students.
Regarding the change from manufacturing to real creation, the key is people and their design. These students have great opportunities while the industry is soaring. But we can see that it would be a long way to become a real fashion designer for the students. I hope that you can practice and educate yourselves constantly after the graduation, which is a new starting point, and can greatly contribute to the industry and the society.

张肇达
著名服装设计师
中国服装设计师协会副主席

我看到，今年同学们的毕业作品很多都充满了激情与创意，很多作品的造型、面料、色彩的运用都很不错，做到了主题突出，思路清晰，内涵丰富，有时代感、有思想、有品位。尤其很多作品开始注重面料创新，这点非常好，因为最有张力的创新设计一定是与现代最新科技元素相结合。同学们在传统的功能和美学的基础上，已具文化艺术素养的哲学思考，在新技术、环保、生态健康、人文关怀等领域的关注，这些层面的思考必将会在未来的设计职业中意义深远。

Zhang Zhaoda
Fashion Designer
Vice President of China Fashion Designer Association

I was amazed by the works of the BIFT graduation collection. The shapes, materials, colors of these wonderful creations impressed me very much. Especially, there were many materials innovation in their design. It is no doubt that this should be encouraged. I believe the most innovative design must have connections with the latest technology. Students should not only study the traditional and aesthetic principles, but also should pay more attention to the new technology, environment protection, eco-system, and humane care. This would play significant roles in all the design careers.

撒宾娜·达曼莎
BETWEEN DESIGN RESEARCH 机构创立者之一

去年三月期间，我在北京偶遇一个老朋友 Nathalie Doucet，旧金山时装艺术基金会主席，她盛情邀请我参加北京服装学院毕业时装秀。13 年来，作为总部设在米兰的设计机构 – Between Design Research– 的创立者之一，我一直都参加在欧洲最负盛名的各个时装学校的毕业演出，但是一直无缘与北服产生接触与交集。

第一次参加北服的毕业演出，我就惊讶于表演的组织，舞台的设计，音乐、影像的质量以及非常了不起的模特表演，这些都令人印象非常深刻。在表演期间，我被各式各样的时装系列深深地感染着。每一个学生都用自己的时装理念打造了一个系列的设计，不断传递着他 / 她的价值观与个性风格。

在我看来，老师们一直做着一项非常有趣的工作，这项工作可以不断加强设计师们的独特观点，以及引导他们打造自己的时装作品，但这些工作并没有干涉到设计师的风格表现。

参加 2012 北服毕业表演非常令人享受，是一个非常棒的经历。学生们的不懈努力和专业知识都在舞台上展现得淋漓尽致，每一件作品的体积感、比例、色彩都非常迷人，缝制的细节也有很高的质量。每一个系列的美，彼此间的融洽与和谐都深深地打动着我。

做的真棒！真希望很快再回来。

Sabrina Damassa
Co-founder of BETWEEN DESIGN RESEARCH

During last March I met by chance in Beijingan an old friend, Nathalie Doucet, President of ARTS OF FASHION FOUNDATION in San Francisco, who kindly invited me to attend the the BIFT final Show.

Since 13 years, as— co founder of a recruitment agency based in Milan – Between Design Research – I have the opportunity to participate the final shows of the most prestigious Fashion Schools in Europe, but till now I hadn't any idea or opinion on the work of the BIFT.

I have been impressed first of all by the organization of the show, the location, the set of the stage, the quality of video and music, as well as the amazing models casting.

Then I have particularly appreciated the variety of the collections.

Every student had created a collection following his own idea of fashion, underlining his/her style and personality.

In my opinion the teachers have done a most interesting work, enhancing designer's peculiarities and personal points of view and helping them to develop their collection without interfere in the style of their projects.

The volumes, proportions and colours of every outlook were charming and the stiching of good quality.

BIFT 2012 Show was an enjoyable and very positive experience. The hard work and goodfashion knowledge of the students have brought to the runway the magic style of different personalities. I was captured by the beauty of the outlooks and the armony and coherence of every single collection.

Well done!! I hope to be back soon again.

坂上 勉
小筱弘子时装品牌创意总监

北服的学生都很优秀，有机会的话，我很想跟他们一起合作。我想说，虽然现在看到了他们所做的作品都很棒，但是在 T 台上面，我看不到这些作品从创意到成品的过程，我看不到他们是如何缝制出来的，看不到他们使用的是什么工艺。如果这些我也能清楚的看到，那就太好了。

Tsutomu Sakagami
Hiroko Koshino Creative Director

BIFT students are very talent and I am quite eager to work with them if possible. Their works are really brilliant. It is nevertheless a pity that I couldn't see how these collections came true from inspiration to a real piece of work, including how they were stitched, and what technology they used. If I could join these process, that would be extremely marvelous.

设计师 / Designer
郭琦 / Guo Qi

指导教师 / Design Advisor
邱佩娜 / Qiu Peina

模特 / Model
那广子 / Na Guangzi

设计师 / Designer
郭琦 / Guo Qi

指导教师 / Design Advisor
邱佩娜 / Qiu Peina

模特 / Model
谢腾 / Xie Teng

设计师 / Designer
丁晓雅 / Ding Xiaoya 李叶晨 / Li Yechen

SEVEN DAYS 2012 "北服杯" 银奖
Silver Award SEVEN DAYS BIFT CUP 2012

指导教师 / Design Advisor
张正学 / Zhang Zhengxue

模特 / Model
陈超 / Chen Chao

设计师 / Designer
刘丽 / Liu Li 赵乃漩 / Zhao Naixuan 刘萌 / Liu Meng

指导教师 / Design Advisor
梁燕 / Liang Yan

模特 / Model
全大川 / Jin Dachuan

设计师 / Designer
郭琦 / Guo Qi

指导教师 / Design Advisor
邱佩娜 / Qiu Peina

设计师 / Designer
郭琦 / Guo Qi

指导教师 / Design Advisor
邱佩娜 / Qiu Peina

模特 / Model
谢腾 / Xie Teng

设计师 / Designer
蔡雨祺 / Cai Yuqi　刘蜜之 / Liu Mizhi

指导教师 / Design Advisor
顾远渊 / Gu Yuanyuan

模特 / Model
李蔚语 / Li Weiyu

设计师 / Designer
蔡雨祺 / Cai Yuqi　刘蜜之 / Liu Mizhi

指导教师 / Design Advisor
顾远渊 / Gu Yuanyuan

设计师 / Designer
刘丽 / Liu Li　赵乃漩 / Zhao Naixuan　刘萌 / Liu Meng

指导教师 / Design Advisor
梁燕 / Liang Yan

模特 / Model
商立强 / Shang Liqiang

设计师 / Designer
刘丽 / Liu Li 赵乃漩 / Zhao Naixuan 刘萌 / Liu Meng

指导教师 / Design Advisor
梁燕 / Liang Yan

模特 / Model
王昱超 / Wang Yuchao

设计师 / Designer
刘丽 / Liu Li   赵乃漩 / Zhao Naixuan   刘萌 / Liu Meng

指导教师 / Design Advisor
梁燕 / Liang Yan

模特 / Model
李桐 / Li Tong

设计师 / Designer
刘丽 / Liu Li 赵乃漩 / Zhao Naixuan 刘萌 / Liu Meng

指导教师 / Design Advisor
梁燕 / Liang Yan

模特 / Model
金大川 / Jin Dachuan

YOU ARE NOT SPECIAL

Section 5
很正经 / So Serious

设计师 / Designer
宋晓琳 / Song Xiaolin 苏玉 / Su Yu 鞠宜杉 / Ju Yishan
SEVEN DAYS 2012 "北服杯" 铜奖
Bronze Award SEVEN DAYS BIFT CUP 2012

指导教师 / Design Advisor
刘卫 / Liu Wei

模特 / Model
李浩一 / Li Haoyi

设计师 / Designer
郑凤 / Zheng Feng

指导教师 / Design Advisor
梁燕 / Liang Yan

模特 / Model
李浩一 / Li Haoyi

设计师 / Designer
宋晓琳 / Song Xiaolin 苏玉 / Su Yu 鞠宜杉 / Ju Yishan

SEVEN DAYS 2012 "北服杯" 铜奖
Bronze Award SEVEN DAYS BIFT CUP 2012

指导教师 / Design Advisor
刘卫 / Liu Wei

模特 / Model
门佳慧 / Men Jiahui

设计师 / Designer
宋晓琳 / Song Xiaolin 苏玉 / Su Yu 鞠宜杉 / Ju Yishan

SEVEN DAYS 2012 "北服杯" 铜奖
Bronze Award SEVEN DAYS BIFT CUP 2012

指导教师 / Design Advisor
刘卫 / Liu Wei

模特 / Model
王能 / Wang Neng

设计师 / Designer
宋晓琳 / Song Xiaolin 苏玉 / Su Yu 鞠宜杉 / Ju Yishan

SEVEN DAYS 2012 "北服杯" 铜奖
Bronze Award SEVEN DAYS BIFT CUP 2012

指导教师 / Design Advisor
刘卫 / Liu Wei

模特 / Model
胡乃月 / Hu Naiyue

设计师 / Designer
陈彦伯 / Chen Yanbo   高燕 / Gao Yan

指导教师 / Design Advisors
尤珈 / You Jia   楚艳 / Chu Yan

模特 / Model
葛晓慧 / Ge Xiaohui

Section 6
不羈 / Unruly Hippie

设计师 / Designer
陈彦伯 / Chen Yanbo 高燕 / Gao Yan

指导教师 / Design Advisors
尤珈 / You Jia 楚艳 / Chu Yan

模特 / Model
陈超 / Chen Chao

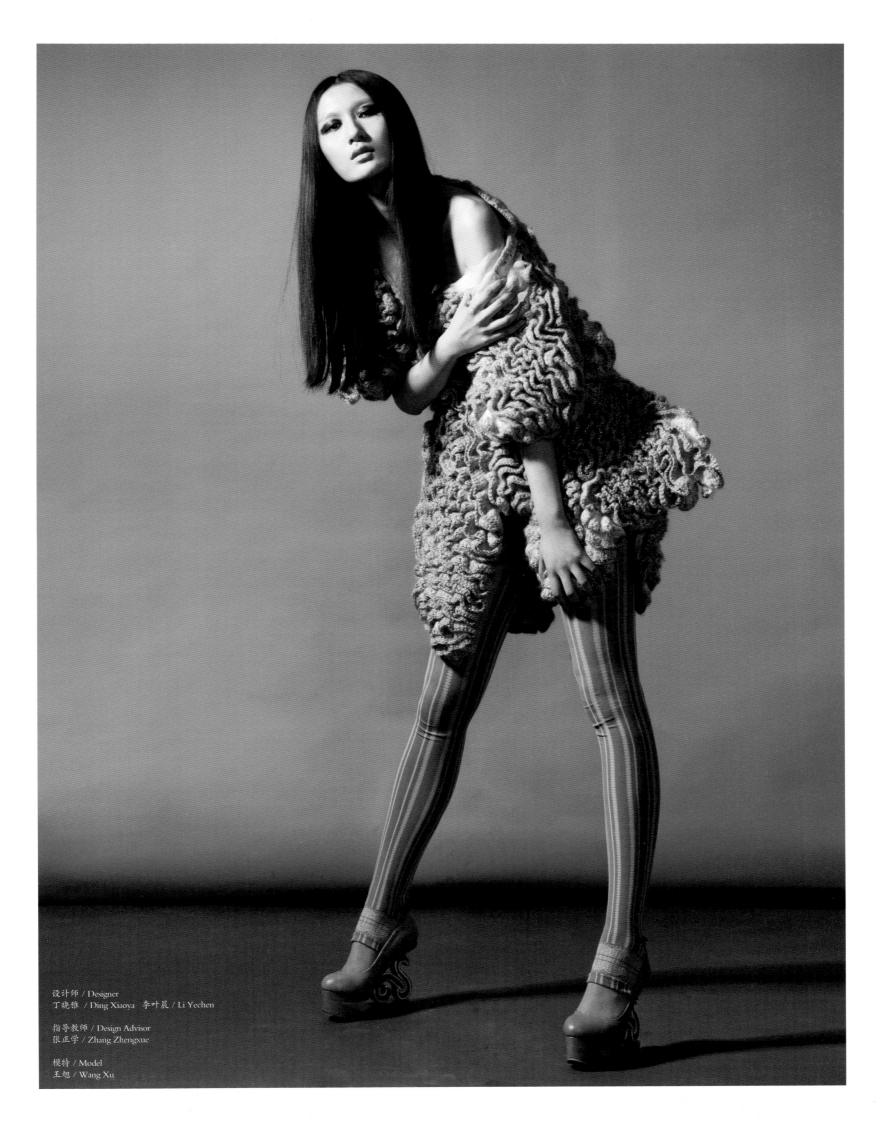

设计师 / Designer
丁晓雅 / Ding Xiaoya 李叶晨 / Li Yechen

指导教师 / Design Advisor
张正学 / Zhang Zhengxue

模特 / Model
王旭 / Wang Xu

设计师 / Designer
陈彦伯 / Chen Yanbo 高燕 / Gao Yan

指导教师 / Design Advisors
尤珈 / You Jia 楚艳 / Chu Yan

模特 / Model
顾燕君 / Gu Yanjun

设计师 / Designer
彭思维 / Peng Siwei

指导教师 / Design Advisor
王昇 / Wang Yi

模特 / Model
李桐 / Li Tong

设计师 / Designer
丁晓雅 / Ding Xiaoya 李叶晨 / Li Yechen

指导教师 / Design Advisor
张正学 / Zhang Zhengxue

模特 / Model
王旭 / Wang Xu

设计师 / Designer
纪兴华 / Ji Xinghua

指导教师 / Design Advisors
谢平 / Xie Ping　张博 / Zhang Bo　邵新艳 / Shao Xinyan

模特 / Model
葛晓慧 / Ge Xiaohui

设计师 / Designer
纪兴华 / Ji Xinghua

指导教师 / Design Advisors
谢平 / Xie Ping　张博 / Zhang Bo　邵新艳 / Shao Xinyan

设计师 / Designer
纪兴华 / Ji Xinghua

指导教师 / Design Advisors
谢平 / Xie Ping 张博 / Zhang Bo 邵新艳 / Shao Xinyan

模特 / Model
印雅琦 / Yin Yaqi

设计师 / Designer
刘一飞 / Liu Yifei 田雅楠 / Tian Yanan

指导教师 / Design Advisors
郭瑞萍 / Guo Ruiping 钟鸣 / Zhong Ming

模特 / Model
胡乃月 / Hu Naiyue

设计师 / Designer
宋晓琳 / Song Xiaolin　苏玉 / Su Yu　鞠宜杉 / Ju Yishan

指导教师 / Design Advisor
刘卫 / Liu Wei

模特 / Model
陈超 / Chen Chao

Section 7
苦旅 / Life is Tramp

设计师 / Designer
张思琦 / Zhang Siqi　王晓莹 / Wang Xiaoying　吕晶 / Lv Jing

指导教师 / Design Advisor
梁燕 / Liang Yan

模特 / Model
金大川 / Jin Dachuan

设计师 / Designer
展晓晴 / Zhan Xiaoqing　韩楚彤 / Han Chutong
谢春燕 / Xie Chunyan　宋亚美 / Song Yamei

指导教师 / Design Advisors
尤珈 / You Jia　杨洁 / Yang Jie

模特 / Model
严恺文 / Yan Kaiwen

设计师 / Designer
展晓晴 / Zhan Xiaoqing   韩楚彤 / Han Chutong
谢春燕 / Xie Chunyan   宋亚美 / Song Yamei

指导教师 / Design Advisors
尤珈 / You Jia   杨洁 / Yang Jie

设计师 / Designer
展晓晴 / Zhan Xiaoqing　韩楚彤 / Han Chutong
谢春燕 / Xie Chunyan　宋亚美 / Song Yamei

指导教师 / Design Advisors
尤珈 / You Jia　杨洁 / Yang Jie

模特 / Model
商立强 / Shang Liqing

设计师 / Designer
宋晓琳 / Song Xiaolin 苏玉 Su Yu 鞠宜杉 / Ju Yishan

指导教师 / Design Advisors
刘卫 / Liu Wei

模特 / Model
王昱超 / Wang Yuchao

Section 8

边缘 / The Edge

设计师 / Designer
李璐 / Li Lu

SEVEN DAYS 2012 "北服杯" 铜奖
Bronze Award SEVEN DAYS BIFT CUP 2012

指导教师 / Design Advisor
楚艳 / Chu Yan

模特 / Model
李浩一 / Li Haoyi

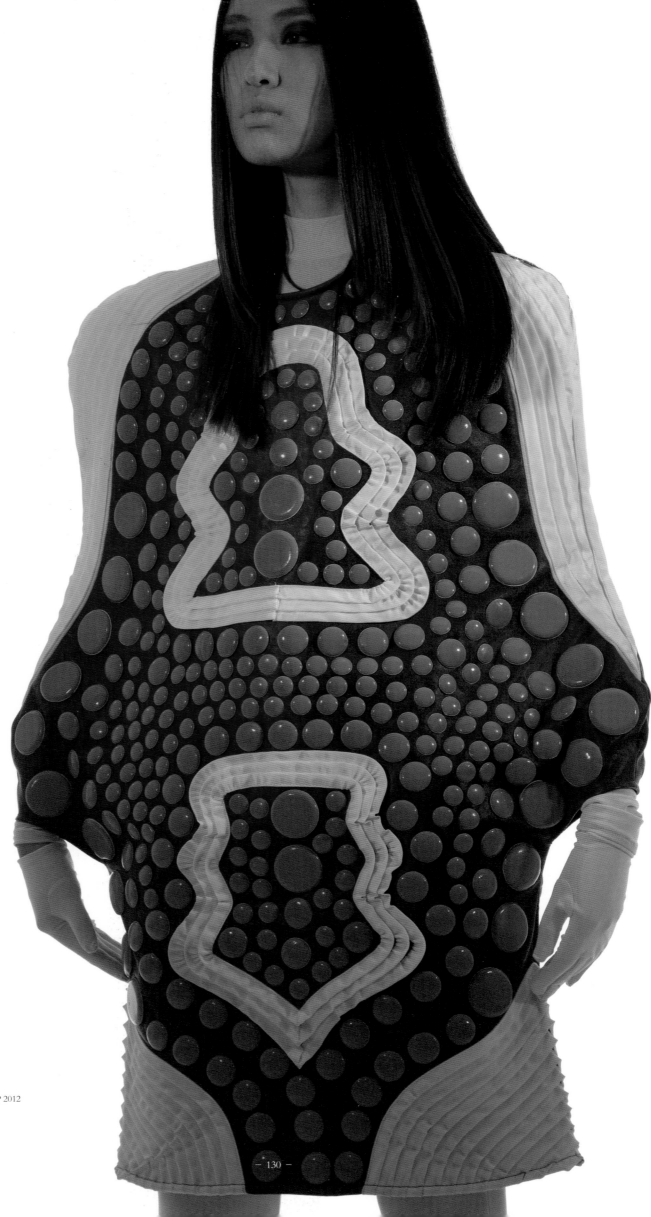

设计师 / Designer
李璐 / Li Lu

SEVEN DAYS 2012 "北服杯" 铜奖
Bronze Award SEVEN DAYS BIFT CUP 2012

指导教师 / Design Advisor
楚艳 / Chu Yan

模特 / Model
顾燕君 / Gu Yanjun

设计师 / Designer
李璐 / Li Lu

SEVEN DAYS 2012 "北服杯" 铜奖
Bronze Award SEVEN DAYS BIFT CUP 2012

指导教师 / Design Advisor
楚艳 / Chu Yan

模特 / Model
韩红盼 / Han Hongpan

设计师 / Designer
李璐 / Li Lu
SEVEN DAYS 2012 "北服杯" 铜奖
Bronze Award SEVEN DAYS BIFT CUP 2012

指导教师 / Design Advisor
楚艳 / Chu Yan

模特 / Model
顾燕君 / Gu Yanjun

2012届毕业设计作品集工作人员名单 / Staff List

顾问 / 刘元凤
总策划 / 赵平 / 王琪 / 谢平
执行总监 / 肖彬
艺术总监 / 苏步
Advisor / Liu Yuanfeng
Proposer / Zhao Ping / Wang Qi / Xie Ping
Executive Director / Xiao Bin
Art Director / Su Bu

摄影 / 苏步
助理摄影 / 张辰
Photographer / Su Bu
Assistant Photographers / Zhang Chen

造型设计 / 肖彬
化妆造型 / 张永晶
Image Director / Xiao Bin
Image Designer / Zhang Yongjing

服装统筹 / 邵新艳
Garment Coordinator / Shao Xinyan

模特统筹 / 刘筱君
Model Coordinator / Liu Xiaojun

平面设计 / 汉禾Design 赵强 高李亚
Graphic Design / HANHE Design Inc. Zhao Qiang / Gao Liya

封面作品 / 周讵燕
Cover Work / Zhou Juyan

翻译 / 黄海峤
Translator / Huang Haiqiao

毕业设计总策划 / 谢平 / 郭瑞萍 / 邵新艳
Graduation Design Directors / Xie Ping/ Guo Ruiping/ Shao Xinyan

指导教师 / （按姓氏笔画排名）
刁杰 / 王丽 / 王羿 / 王群山 / 王媛媛 / 尤珈 / 兰岚 / 刘卫 / 刘娟 / 孙雪飞 / 杜冰冰 / 李玮琦 / 苏步 / 肖彬 / 张博 / 张正学
邱佩娜 / 邵新艳 / 杨洁 / 邹游 / 周邵恩 / 赵明 / 钟鸣 / 郑嵘 / 顾远渊 / 常卫民 / 郭瑞萍 / 黄洪源 / 梁燕 / 谢平 / 楚艳
Design Advisors
Diao Jie / Wang Li / Wang Yi / Wang Qunshan / Wang Yuanyuan / You Jia / Lan Lan / Liu Wei / Liu Juan / Sun Xuefei / Du Bingbing / Li Weiqi
Su Bu / Xiao Bin / Zhang Bo / Zhang Zhengxue / Qiu Peina / Shao Xinyan / Yang Jie / Zou You / Zhou Shaoen / Zhao Ming / Zhong Ming
Zheng Rong / Gu Yuanyuan / Chang Weimin / Guo Ruiping / Huang Hongyuan / Liang Yan / Chu Yan

"北服·SEVENDAYS"优秀毕业设计奖学金获奖学生及作品名称

金奖：张晓田《融化的冰山 Melting iceberg》

银奖：李佳佩《THE END IS THE BEGINNING》
银奖：丁晓雅 李叶晨《游弋》

铜奖：赵荣峰 刘建凯 胡文邦《凝聚》
铜奖：宋晓琳 苏玉 鞠宜杉《很正经》
铜奖：李璐《边缘 bpd》

优秀奖：刘丽 赵乃漩 刘萌《噬》
优秀奖：周讵燕《逆向》
优秀奖：李胜男 张健杨 雨心《雕梁绣意》
优秀奖：蔡雨祺 刘蜜之《First frost 霜降》
优秀奖：刘一飞 田雅楠《味道》
优秀奖：靳云英 陈小姣 王雪桦 王晓旭《劣迹斑斑》/《破晓》
优秀奖：胡文科 胡羚群 王元 吕熹《WOW》
优秀奖：杨爽 马华远《潋》
优秀奖：郭琦《CHANGE》

最佳工艺奖：张晓田《融化的冰山 Melting iceberg》

最佳表演奖：赵晨池

奖学金提供：上海世芬笛施服饰商贸有限公司

鸣谢（排名不分先后）/ Thanks

上海世芬笛施服饰商贸有限公司
北京汇博天慧文化传媒有限公司
上海索雅时装有限公司
广州市质品服饰有限公司
广东东莞超盈纺织有限公司
Saga Furs Oyj 世家皮草
CCTV《创意星空》栏目
浙江新澳纺织股份有限公司
江苏金龙科技股份有限公司
江苏鹿港科技股份有限公司
上海协大国际贸易有限公司
上海银之川金银线有限公司
广东伽懋毛织时装有限公司
汕头市天辉毛织制品有限公司
汶上如意天容纺织有限公司
无锡富士时装有限公司
三发成（上海）国际贸易有限公司
宇宙星国际商业管理（北京）有限公司
北京奥索克体育用品有限公司
广东小猪班纳服饰股份有限公司
广州市卡宾服饰有限公司
山东南山纺织服装有限公司

模特 / Models

赵晨池 / 2006、2007、2010 年度十佳职业模特
Zhao Chenchi / 2006/2007/2010 China Top 10 Professional Fashion Models
王梦雅 / 2011 年度十佳职业模特
Wang Mengya / 2011 China Top 10 Professional Fashion Models
李蔚语 / 2010 年第十一届 CCTV 模特电视大赛 冠军
Li Weiyu / Champion of the 11th CCTV Model Contest in 2010
那广子 / 2010 年第十一届 CCTV 模特电视大赛 季军
Na Guangzi / Second Runner-up of 11th CCTV Model Contest in 2010
李浩一 / 2010 年亚洲超模大奖最佳新人奖
Li Haoyi / The Best New Model of 2010 Asian Supermodel Contest
蔡浩 / 2011 第十七届中国模特之星大赛 冠军
Cai Hao / Champion of China Model Star Contest 2011
赵冰清 / 2011 中国职业模特选拔大赛 冠军
Zhao Bingqing / Champion of 2010 China Professional Model Contest
王旭 / 2008 第 14 届中国模特之星大赛 亚军
Wang Xu / Runner-up of China Model Star Contest 2008
李姿含 / 2011 环球小姐中国赛区 亚军
Li Zihan / Runner-up of Miss Universe China Contest 2011
葛晓慧 / 2009 中国模特新面孔选拔大赛 冠军
Ge Xiaohui / Champion of 2009 New Faces China Model Contest
贾亦真 / 2010 中国模特新面孔选拔大赛 亚军
Jia Yizhen / Runner-up of 2010 New Faces China Model Contest
陈超 / 2010 年中国新丝路模特大赛 亚军
Chen Chao / Runner-up of 2010 New Silk Road China Model Contest
金大川 / 2011 中国模特之星大赛 季军
Jin Dachuan / Second Runner-up of China Model Star Contest 2011
徐文晰 / 2011 世界华人美后大赛 冠军
Xu Wenxi / Champion of Miss Chinese World 2011
严恺文 / 2010 中国模特新面孔选拔大赛 亚军
Yan Kaiwen / Runner-up of New Faces China Model Contest in 2010
李博文 / 2011 中国模特新面孔选拔大赛 冠军
Li Bowen / Champion of New Faces China Model Contest in 2011
李沛霖 / 2012 亚洲新人模特大赛 冠军
Li Peilin / Champion of 2012 Asia New Face Model Contest
李梦涵 / 2011 北京下一位超模大赛 亚军
Li Menghan / Runner-up of Beijing Next Top Model Contest